OIL
SPILL

OIL SPILL

CHRISTOPHER LAMPTON

THE MILLBROOK PRESS
BROOKFIELD, CT
A DISASTER! BOOK

Cover photograph courtesy of Magnum Photos © Paul Fusco

Photographs courtesy of Magnum Photos : pp. 9 (© Paul Fusco),
36 (© Larry Towell); Pennsylvania Historical & Museum Commission's
Drake Well Museum, Titusville, Pa.: pp. 12 (top), 19; Superstock:
pp. 12, 16, 20–21; Wide World Photos: pp. 26, 27, 33; Photo
Researchers: p. 31 (M.B. Duda); Greenpeace: p. 38 (Merjenburgh);
U.S. Department of Energy: p. 41. Illustration by Pat Scully.

Library of Congress Cataloging-in-Publication Data

Lampton, Christopher.
Oil spill / Christopher Lampton.

p. cm.—(A Disaster! book)
Includes bibliographical references and index.
Summary: Describes how and why oil spills happen, the damage such
accidents can do to the environment, and methods used to clean up spills.
ISBN 1-56294-071-6 (lib. bdg.)
1. Oil spills—Environmental aspects—Juvenile literature. 2. Oil
pollution of water—Environmental aspects—Juvenile literature. [1.
Oil spills. 2. Petroleum industry and trade.] I. Title. II. Series:
Lampton, Christopher. Disaster! book.
TD427.P4L36 1992
628.5′2—dc20 91-43565 CIP AC

CONTENTS

THE EXXON VALDEZ DISASTER

The ship was headed for disaster—and nobody on board knew it.

The *Exxon Valdez* was an oil tanker. It carried oil from Alaska to the lower forty-eight states. Its journey started in Valdez (pronounced val-DEEZ), the town from which the ship took its name. The Alaskan oil pipeline, which carries oil from the vast oil fields of Alaska to the sea, ends in Valdez. It was here that the *Exxon Valdez* filled its huge tanks with oil, then sailed out of the bay toward the open ocean.

But before it could reach the ocean, the *Exxon Valdez* had to pass through some treacherous waters. Valdez Bay opens into a body of water known as Prince William Sound. On the night of March 23, 1989, Prince William Sound was filled with icebergs.

The easiest way to sail a ship in such waters is to slow down to a crawl and creep past the danger spots. But the crew of the *Exxon Valdez* didn't want to slow down. The longer it took them to deliver the oil, the more it would cost the Exxon Corporation, which owned the ship. So they steamed through the ice at nearly full speed.

Still, there should have been no problem. Most of the crew had experience with ice and knew how to steer around it. Yet, somehow, the ship got off course, drifting close to an underwater reef next to Bligh Island, which is on the eastern side of the sound.

Fortunately, the reef was well marked by buoys, and there should have been plenty of room to get around it. Unfortunately, there was an iceberg in the way.

The captain ordered the ship's helmsman to steer between the iceberg and the reef. The young man tried to execute the captain's orders, but there just wasn't enough room.

Shortly after midnight on March 24, the *Exxon Valdez* hit the reef. A shudder passed through the ship. Eight gaping holes ripped open in its side, and oil began to pour out. An oil spill was in progress.

Eleven million gallons of oil poured from the *Exxon Valdez*'s tanks into Prince William Sound. This was only one fifth of all the oil that the ship was carrying, but it was enough to make the incident a major disaster.

Oil and water don't mix, the old saying goes. Oil doesn't mix with things that live in the water, either. The oil from the *Exxon Valdez* killed fish by the thousands. Dead sea birds, their feathers covered with oil, washed up on the beaches in huge numbers. The beaches themselves were soon covered with a black, tarlike substance. Attempts to clean up the results of the *Exxon Valdez* oil spill were still going on years after it happened.

Within hours, news of the *Exxon Valdez* oil spill had been broadcast all over the world. The public was outraged. The idea of messing up the clear waters and clean white beaches of Prince

The Exxon Valdez *anchored for repairs off the Alaskan coast. The red object surrounding it is a containment boom.*

William Sound angered people the world over. Many people loudly denounced the Exxon Corporation and demanded that it be punished for the spill.

How did an accident such as the *Exxon Valdez* oil spill happen? Why was a ship filled with oil allowed to recklessly sail through a body of water containing reefs and icebergs? The answer lies in the unique value of the substance that the ship was carrying—*petroleum,* better known simply as oil.

OIL MAKES THE WORLD GO 'ROUND

Oil is one of the most important substances on earth. It provides much of the energy that makes our technological society go. Oil in the form of gasoline makes our cars run. Oil heats many of our homes. Oil can be found in plastic and in clothing. In centuries past, oil was even used in medicine.

Oil is needed all over the world. Unfortunately, there are only a relatively few places where oil can be found in large quantities. Some method is needed to get oil from the few places where it is found to the many places where it is needed. This is why tankers such as the *Exxon Valdez* exist.

But when oil is moved around in large quantities, some of it is almost certainly going to get spilled. Most oil spills are quite small. In fact, there are probably oil stains on the street not far from where you live, where oil spilled out of the engine of a car. Even small oil spills such as these can damage our environment. But the worst oil

11

spills usually occur at sea, when ships such as the *Exxon Valdez* are damaged and release their valuable cargo into the water.

In this book, we'll talk about some of the worst oil spills of all time. But we'll also talk about oil itself—what it is, where it comes from, and why it's so valuable.

In the 1860s, oil was transported to markets by "flatboating" on rivers and creeks, as shown above, left. Today, supertankers like the one at left carry oil around the world.

WHERE OIL COMES FROM

When we talk about the kind of oil that is carried in tankers such as the *Exxon Valdez,* we are actually talking about petroleum. Where does petroleum come from? Usually, it comes from deep in the ground, where it collects inside the very rocks of the earth itself. In fact, the word *petroleum* comes from Latin words meaning "rock oil."

How does this petroleum get into the rocks? And where does it get the energy that makes our cars run and our furnaces burn? It got into the rocks when the rocks were forming. And it gets its energy from the sun.

Look out your window, and you'll probably see lots of living organisms. Unless you are in the middle of a large city, most of these living organisms will be green. We call these green organisms *plants.*

Unlike other living organisms, plants are not very active. They spend most of the day sitting in the sun, apparently doing nothing.

But don't be fooled. When plants appear to be doing nothing, they are actually gathering energy from the sun.

This is how the sun's energy passes from one living organism to another: First, it is absorbed by plants directly from the sun. Then, the plants are eaten by animals. Then, the animals are eaten by other animals. This process is known as the *food chain* or *food web.* If it weren't for the food chain, living organisms wouldn't have the energy they need to move and grow.

Not all animals are eaten by other animals, of course. When an animal is buried beneath the ground, for instance, the carbohydrates in its body return to the soil. But what happens when a living organism dies in water?

This has happened many times in the history of the earth. In fact, all sorts of things fall into the water. Eventually they settle to the bottom (as sediment), where still more things settle on top of them. Slowly, the weight of all the sediment squeezes everything together until rock is formed.

The type of rock formed at the bottom of rivers and oceans is called *sedimentary rock.* After many years, when the water dries up, this sedimentary rock can come back to the surface to form the ground that we walk on. But even this seemingly solid rock can contain tiny holes, called pores, as we shall see.

What happens to the living organisms that die in the water or fall into the water after they die? Do they become part of the sedimentary rock?

Yes, they do. Over the course of millions of years, the weight of the overlying rocks squeezes these dead plants and animals until they become a thick, oily substance. They become petroleum. And this petroleum still contains the solar energy that these plants and animals received from the food chain.

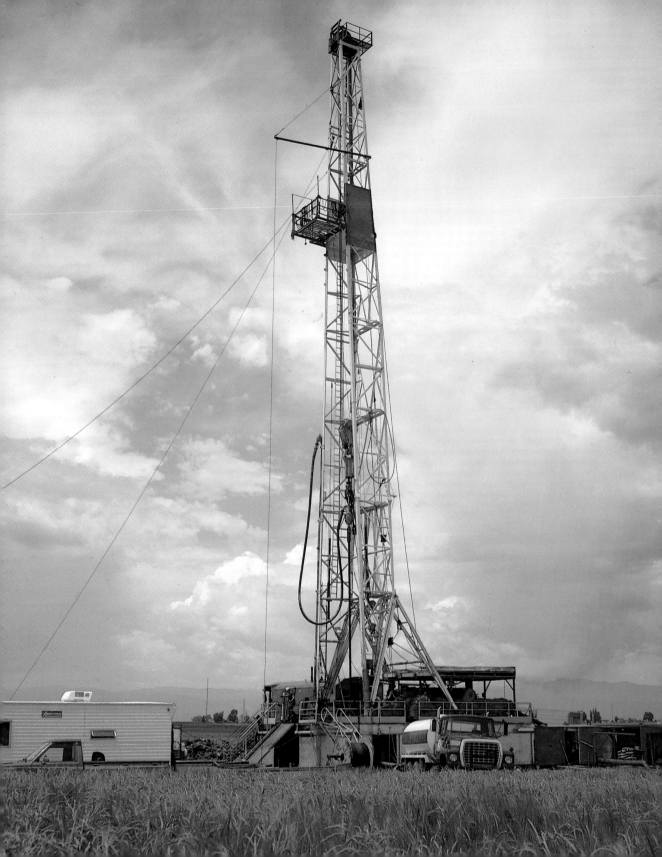

This is how the petroleum gets into the rocks. Over the years, it seeps slowly through the tiny pores in the rocks. Sometimes large amounts of petroleum gather together into large underground pools. Scientists are still not sure what causes the oil to move through the rocks and gather in pools.

How do we get the petroleum back out of the ground? The most common method is to build an *oil well.*

Oil can be found under the ocean or under dry land. This oil drill pumps petroleum from under a wheat field in Colorado.

THE PETROLEUM INDUSTRY

In 1859, an American inventor named Edwin L. Drake had a bright idea. Oil was in demand as fuel for lamps. Drake knew that oil could be found underground. So, he reasoned, why not build a drill that could dig into the earth and pump the oil to the surface? Thus was born the oil well!

Drake dug his first well outside of Titusville, Pennsylvania. It was 69 feet (20.7 m) deep. The oil that it pumped was sold by the Seneca Oil Company, for which Drake worked.

The type of oil that comes out of a well is called *crude oil.* Crude oil isn't usable until it has been processed. The first step in processing crude oil is to heat it until it begins to boil.

As the oil is heated, the various parts of the oil, called *fractions,* come to a boil at different temperatures. When a fraction boils, it turns into a gas. The gases released in this way can be trapped and turned back into liquids.

As the temperature goes up, the first fraction to boil off is gasoline, which we use as fuel in our cars. Then, as the temperature

Edwin L. Drake built the first oil well in
Titusville, Pennsylvania, in 1859. Here
he is shown (right) in front of his well.

This Arco oil refinery in Texas City, Texas, processes crude oil into gasoline, kerosene, and other usable forms.

continues to rise, kerosene boils off. Finally, at the highest temperatures diesel and heating oils boil off. After the boiling is complete, a sticky residue is left behind. This residue is used to make asphalt, a road paving substance.

Petroleum is used in many ways that you might not expect. It is an ingredient in many types of plastic, for instance, and it is even used in the production of synthetic fibers for clothing. It is used in the manufacture of some soaps and insecticides and is also found in rubber tires and vinyl phonograph records.

As you can see, petroleum is useful stuff. It's so useful that an

entire industry has developed around it. Some of the largest corporations in the world are involved in drilling for, processing, and selling oil.

Most of the oil in the world is used by large nations, including the United States and the Soviet Union. But the oil itself often comes from smaller countries. A country with a large supply of oil has considerable political power internationally. In the 1960s, several of the oil-producing countries formed the Organization of Petroleum Exporting Countries (OPEC, for short). When the OPEC nations decided to raise the price of oil, it caused a worldwide oil crisis.

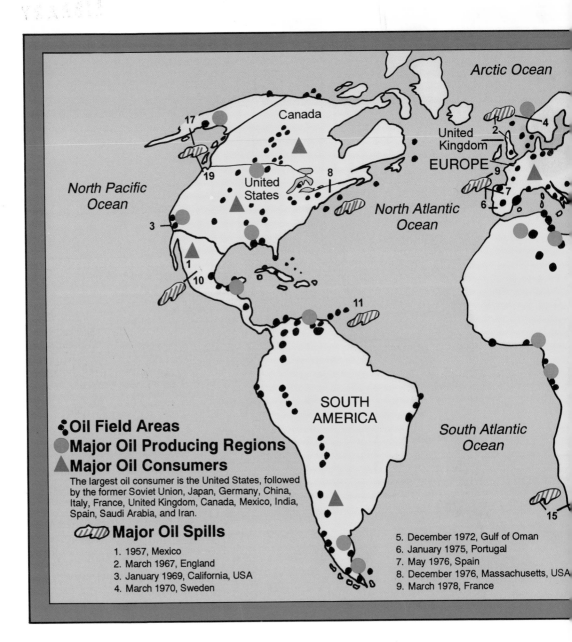

Oil Field Areas

Major Oil Producing Regions

Major Oil Consumers

The largest oil consumer is the United States, followed by the former Soviet Union, Japan, Germany, China, Italy, France, United Kingdom, Canada, Mexico, India, Spain, Saudi Arabia, and Iran.

Major Oil Spills

1. 1957, Mexico
2. March 1967, England
3. January 1969, California, USA
4. March 1970, Sweden
5. December 1972, Gulf of Oman
6. January 1975, Portugal
7. May 1976, Spain
8. December 1976, Massachusetts, USA
9. March 1978, France

Map labels: Arctic Ocean, Canada, United Kingdom, EUROPE, North Pacific Ocean, United States, North Atlantic Ocean, SOUTH AMERICA, South Atlantic Ocean

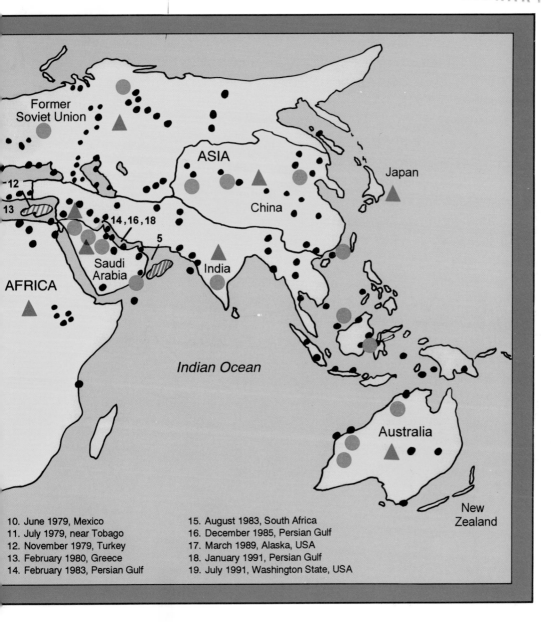

Former
Soviet Union

ASIA

Japan

China

−12

13

14, 16, 18

5

Saudi
Arabia

India

AFRICA

Indian Ocean

Australia

New
Zealand

10. June 1979, Mexico
11. July 1979, near Tobago
12. November 1979, Turkey
13. February 1980, Greece
14. February 1983, Persian Gulf

15. August 1983, South Africa
16. December 1985, Persian Gulf
17. March 1989, Alaska, USA
18. January 1991, Persian Gulf
19. July 1991, Washington State, USA

23

The world's supply of petroleum is limited. Eventually, it will run out. Because it takes millions of years for oil to form in sedimentary rock, we can't expect a new supply of petroleum to appear any time soon. For that reason, most experts believe that we must eventually find new kinds of fuels to run our automobiles and other petroleum-powered vehicles.

Huge amounts of petroleum are transported around the globe every day. To transport all this oil, a new type of ship has been developed: the *supertanker.* Supertankers, which are almost like floating cities, can each carry more than 200,000 tons of oil. Some supertankers carry more than 500,000 tons of oil! Of course, smaller tankers also carry oil. All of these ships are essentially floating oil tanks.

As the amount of oil carried has grown larger and larger, the chances that some of that oil will get spilled have also grown larger. And it's common for one tanker to transfer its oil to a second tanker in the middle of the ocean, which makes a spill even more likely. Inevitably, some of this oil gets into the water. Usually, the amount of oil spilled is small, but sometimes large spills occur.

And when something goes very wrong—when an oil tanker crashes, as the *Exxon Valdez* did—the oil spill can be very large indeed.

THE LARGEST SPILLS OF ALL TIME

The general public didn't pay much attention to oil spills until the late 1960s. Then, in 1967, one of the first supertankers, the *Torrey Canyon,* ran aground on Pollard Rock off the coast of England. A simple error—a lever left in the wrong position—caused the helmsman to lose control of the ship. When the captain realized that there was a problem, he tried to stop the ship. But it was too late. Once a supertanker starts moving, it takes a long time to bring it to a stop. Getting it to turn is almost as hard.

As a result, the *Torrey Canyon* plowed into Pollard Rock at a high speed. More than 500,000 barrels of oil spilled out of the ship and into the sea. Not only did this oil wash up on the coast of England, but it also reached neighboring countries, including France and Spain.

The *Torrey Canyon* disaster made headlines around the world. Legislation governing the way in which oil tankers carried oil was passed in the aftermath of the crash.

The Torrey Canyon *spill.*

But it was not to be the last major oil spill. Far from it. In addition to the *Exxon Valdez* spill, there have been several other major oil spills in the decades since the *Torrey Canyon* disaster.

■ One of the two largest oil spills in history occurred in 1979—and it didn't involve an oil tanker. The Ixtoc I oil well, located in the middle of the Gulf of Mexico, accidentally sank its drill into an underwater oil reservoir. The huge pool of oil came bursting up into the water, spilling 140 million gallons of oil directly into the gulf. When oil comes bursting out of a well in an uncontrolled fashion like this, it's called a *blowout.*

An equally large spill occurred in 1991. During the Persian Gulf War, the Iraqi army is believed to have dumped hundreds of thousands of gallons of oil from a loading dock in Kuwait straight into the Persian Gulf. To make matters worse, they also may have dumped the contents of several supertankers into the gulf. The result was an oil spill comparable to the Ixtoc I disaster. In fact, this oil spill was at first believed to dwarf all other oil spills in history.

Crude oil washed up on a beach at Khafji, Saudi Arabia, in January 1991, during the Gulf War. Iraqi forces were suspected of deliberately releasing oil into the Persian Gulf.

Fortunately, the first estimates of the size of the spill turned out to be wrong. Still, this was a nasty oil spill indeed, all the more so because it was deliberately done. Iraqi soldiers later set fire to a portion of the oil slick.

■   England was struck again by oil disaster in 1978, only nine years after the *Torrey Canyon* spill. This time, the supertanker *Amoco Cadiz* ran aground in the English Channel. Nearly 70 million gallons of oil were dumped into the waters—more than six times as much as in the *Exxon Valdez* spill.

■   In 1969 there was an oil well blowout similar to the one that later occurred at Ixtoc I. This time, the oil well known simply as Platform A in the Santa Barbara (California) Channel released four million tons of oil into the water.

■   One of the earliest major spills took place in 1957, when an oil tanker called the *Tampico Maru* crashed into the coast of Mexico. It released 2.7 million gallons of oil into the sea.

■   On July 22, 1991, a Japanese fish-processing ship collided with a Chinese freighter off the coast of Washington, in the northwestern United States. One person was killed in the accident, and 100,000 gallons of oil stored on the Japanese ship poured into the Pacific Ocean. By the standards of major oil spills, this was not particularly large. By comparison the *Exxon Valdez* accident released 11 *million* gallons of oil. But the accident occurred near a wilderness beach where a large number of fledgling birds had recently been born. Many of these fledglings were soon found washed up on Washington beaches, dead. And the beaches are so remote from towns and highways that rescue workers had a difficult time reaching the site to save the birds. Many rescue workers had to be flown in by helicopter.

Of course, as we saw earlier, not all oil spills are major disasters. The vast majority of oil spills are quite small. And yet they add up. For instance, if you dump the oil from your lawn mower onto the grass in your front lawn, you may think nobody will be harmed by the spill. But that small "oil spill" will soak into the ground and mingle with the groundwater that flows through the underlying rocks and clay. From there, it can harm living organisms over quite a large distance. And an oil spill in the street (or on your lawn) can harm birds and other animals that come to investigate it.

In a real sense, there is no such thing as a minor oil spill. Every oil spill is a potential disaster.

The *Exxon Valdez* and *Amoco Cadiz* oil spills were nowhere near as large as the Ixtoc I and Persian Gulf spills. In many ways, however, they may have been even more disastrous. What matters in an oil spill is not so much the amount of oil that is spilled as what happens to that oil after it is spilled. It is in the aftermath of an oil spill that the *real* disaster occurs.

THE
AFTERMATH

Anyone who has ever gazed out over a body of water knows how beautiful the sea can be. When the sun is rising or setting over the water and the ocean seems to be on fire with its reflection, it's hard to imagine how any other place on earth could be quite as lovely. The thought of all this beauty being spoiled by thousands of gallons of thick, black oil is enough to enrage the most reasonable person.

And yet, an oil spill can be quite beautiful in its fashion. The oil slick left behind by the spill can shimmer with all the colors of the rainbow when the light catches it in the right way. You can see this effect for yourself by looking at the sun striking oil puddles on the street outside your house.

The effect of this oil on the things that live in the water is something else. Oil can be devastating to the ecology of the ocean, that is, to the way the living things in the ocean interrelate and sustain themselves in the fragile ocean environment. Oil can kill living things, can disrupt the food chain, and can ruin beaches.

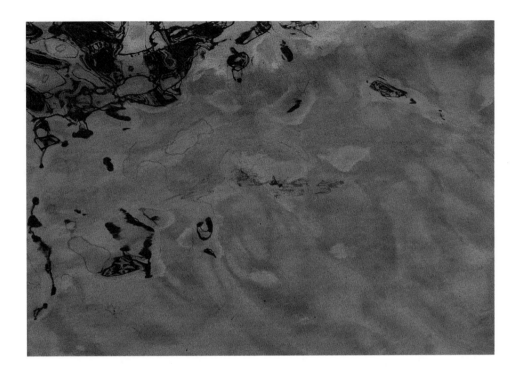

Although deadly to wildlife, oil slicks can be quite beautiful, shimmering with bright colors in the light.

Fortunately, nature itself can take care of a large percentage of any oil spill. When an oil spill occurs, the oil floats on the surface of the water, like a blanket. Some of this oil then evaporates. That is, it turns into a gas and floats away in the air. The motion of the water and the wind breaks up much of the oil slick that is left behind. Some of the oil sinks into the water, settling to the bottom of the sea. (Whether this sunken oil causes further damage depends on where the spill took place. Most of the living organisms on the sea bottom live close to shore. Thus, oil that sinks to the bottom far from shore causes little damage.) Tiny living organisms, a type of bacteria, actually eat part of the oil. As a result, most of an oil spill disappears within a few days after the spill takes place.

Still, roughly one fourth of the spill is still left behind after these processes take place. If the spill was large enough, this is enough to cause considerable damage.

The most obvious way in which oil can harm the animals that live in or near the water is by covering them with oil. Fish-eating birds that live near the sea can easily become dipped in the oil. And, when their feathers are covered with oil, the birds commonly find that they are unable to fly—and therefore unable to find food. The oil also disturbs the birds' metabolism, the bodily processes that keep them alive. A bird covered with oil will usually die fairly quickly.

Obviously, fish are also affected by oil spills, as much by the disruption of their habitat as by the poisonous effects of the oil itself. The oil can destroy fish eggs, for instance, or prevent fish from migrating.

Perhaps the worst damage of all is to the food chain. We discussed the food chain earlier in this book. It is the way in which the energy of the sun reaches all living creatures. Many sea creatures obtain their energy by eating microscopic organisms called plankton, which live in the water. If oil kills the plankton, then the animals that eat the plankton will also die. And then the animals that eat the animals that eat the plankton will die.

Oil is also bad for the fishing industry. Fish that eat plankton full of oil—if this doesn't kill them—will take the oil into their own bodies. If we catch and eat these fish, we'll actually be eating the oil that was spilled into the water. Nobody is sure whether or not it is harmful to human beings to eat organisms tainted with oil. But

*A thick layer of crude oil coats the feathers
of this cormorant as it stands in oily water.
Birds covered with oil usually die quickly.*

there is some evidence that oil can cause cancer. Thus, an oil spill can ruin the livelihood of fishers and others who make their living from the sea. There is little market value in seafood tainted with crude oil.

Beaches covered with oil are an unsightly mess. And the living organisms on the beach are as much at risk as the organisms in the sea. Birds can become coated with oil on the beach. And plants covered with oil cannot soak up energy from the sun. Inevitably, they die.

The severity of an oil spill depends to a large extent on where it takes place. If it takes place in the middle of the ocean, where relatively few living organisms are to be found, it will have a relatively small impact. But if it takes place close to shore, many thousands of organisms can be affected, and the results can be devastating.

The Ixtoc I oil spill, although one of the two largest in history, caused relatively little damage because it took place in the middle of the Gulf of Mexico. Most of the oil simply sank back into the ocean, where it did little or no noticeable harm. The *Amoco Cadiz* and *Exxon Valdez* spills, on the other hand, were a lot smaller but also a lot more disastrous, because they took place close to shore. The *Amoco Cadiz* spill may well have been the worst oil disaster in history, in terms of its effect on the environment. Damage from the spill was still visible a decade after it happened. It's too early to say whether the *Exxon Valdez* spill (or the Persian Gulf spill) will prove even more disastrous.

FIGHTING
BACK

As long as oil is used, oil spills will keep happening. In fact, even if there were no oil tankers transporting the oil, the two worst spills of all time would still have taken place. The Ixtoc I spill and the Persian Gulf spill didn't involve tankers.

Therefore, there must be a way to clean up spills after they happen. And indeed, a number of methods have been developed to fight spills. Unfortunately, none of these methods is perfect. Damage from oil spills still results, no matter how hard we fight back. Still, there is a lot that can be done to minimize the damage that takes place.

One of the most common tools for fighting oil spills is the *boom*. A boom is a kind of floating fence. It can be placed around an oil spill to keep it from spreading. Once a boom is in place, the oil can be set on fire until most of it has burned away.

A boom being used to skim oil from the waters of Prince William Sound, Alaska, following the Exxon oil spill.

Some booms are inflatable, like giant balloons. The workers trying to contain the oil spill simply inflate the boom and loop it around the oil, like a giant lasso.

Unfortunately, booms aren't perfect. If the oil slick is thin, it can slip underneath the boom. If the water is moving rapidly, the boom may be simply tossed aside.

Even if the oil can't be contained by a boom, there are other methods of cleaning it up. For instance, huge machines, like giant vacuum cleaners, can suction up the oil from the water. Devices for sucking oil off the surface of the water are known as *skimmers* and *pumpers.* Skimming and pumping the oil into large tanks allows the oil to be used again, which saves a lot of money for the company that owns the oil.

Detergents can be used to break up the oil and force it to sink. We said earlier that oil and water don't mix. Because of the difference in the weight and consistency of the two substances, oil tends to sit on top of water in a solid mass. But detergent can break the oil into tiny droplets that sink in the water and are dispersed by the currents. Of course, the oil companies don't like to use detergents because they are unable to recover the oil and process it. And the detergents aren't necessarily good for the water either.

One of the newest, most ingenious techniques is to use natural bacteria to eat the oil. Natural processes such as wind, waves, and bacteria can disintegrate much of an oil spill. To help these processes along, some oil spill cleanup experts have suggested that large colonies of bacteria be dumped on a spill to devour it before it can do much damage. In fact, special strains of bacteria have been developed to eat large amounts of oil.

So far, however, this method has not been used much, because the bacteria devour the oil at a very slow pace. After the *Exxon Valdez* spill in Alaska, the Exxon Corporation attempted to use bacteria to clean up beaches fouled by the oil. However, bacteria eat even more slowly in cold weather than they do in warm, and the weather in Alaska in March is quite cold. The company overcame this problem, to some extent, by putting a layer of fertilizer on top of the bacteria. The fertilizer caused the bacteria to mul-

tiply. The individual bacteria still ate as slowly as before, but now there were more of them doing the eating. This is a technique that may be more widely used in cleaning up future oil spills.

Saving the oil-coated wildlife left behind by an oil spill is a major job in itself. You've probably seen pictures of rescue workers cleaning birds after an oil spill such as the *Exxon Valdez* accident. This is a difficult job, in some ways more difficult than cleaning up the spill itself. After all, birds and otters and other animals are not going to cooperate with those trying to help, since they don't know that the humans are trying to save their lives. Just catching the animals in order to clean them is a major task. And cleaning oil out of fur and feathers is very difficult.

The care and dedication of rescue workers at a rehabilitation center in Valdez saved the lives of these sea otters. Many others were not so fortunate.

STOP THAT SPILL

Can we prevent oil spills in the future? Probably not—unless we decide to live without any oil at all.

Through energy conservation, however, we can cut down on the amount of oil we use. You can help in this endeavor. For instance, you can encourage your family not to make unnecessary car trips and to use public transportation whenever possible. You can turn off lights that are not in use. (The electricity that makes the lights burn is sometimes generated by burning oil.) You can help educate your friends and neighbors about the need to use less oil. And you can write letters to your representatives in Congress asking them to vote for measures supporting energy conservation.

Another way to cut down on oil spills is to encourage the use of alternative forms of energy generation. Scientists are looking for fuels that can replace gasoline in powering cars and for new methods of producing electricity, such as solar energy. We can also learn to use oil more efficiently—to get more work out of every gallon burned.

Less oil use means fewer oil spills, and scientists are working to find alternative energy sources. This is a solar-power test facility in New Mexico.

It's unlikely, though, that people will stop using oil completely until the supplies run out. In the meantime, we must do our best to make oil spills as unlikely as possible. One method that might lower the risk of oil spills is to give supertankers double bottoms. This means simply building a second bottom inside the existing bottom of the ship's hull. Both bottoms would have to break open before the oil could escape the ship.

But there are problems with this method. For instance, water trapped between the two bottoms could cause a ship to sink more quickly than would a ship with a single bottom. This could cause even more oil to be released during a spill. And it's possible for gases to build up in the gap between the two bottoms, possibly resulting in an explosion that could shatter both of the ship's hulls. For these reasons, there's no general agreement as to whether a double bottom would prevent disastrous spills—or increase them.

We said earlier that the *Torrey Canyon* disaster resulted in legislation concerning the transportation of oil. This legislation, which requires oil transporters to develop plans for dealing with oil spills, has probably done a lot to prevent oil spills from happening. It requires that all oil spills be reported immediately and cleaned up as soon as possible. Fines are levied on anyone who spills more than a certain amount of oil.

Punishing those who spill oil, with fines and jail sentences, is a great deterrent. But it can't stop spills. As long as there is oil, there will be oil spills. We can just hope that they happen rarely and that there will always be dedicated people to clean them up!

GLOSSARY

blowout—oil bursting out of a well in an uncontrolled fashion.

boom—device, similar to a floating fence, for trapping an oil spill.

crude oil—the form in which oil is extracted from the ground.

food chain—the way in which the sun's energy is passed from one living organism to another. Also called the *food web.*

fractions—the oil products that separate out when crude oil is boiled.

oil well—drilling apparatus used to bring oil to the surface from beneath the ground.

petroleum—oily substance that formed from the remains of living organisms trapped in sedimentary rock over millions of years. Usually measured in gallons, tons, or barrels.

plankton—microscopic plant life in the ocean.

plants—living organisms that obtain their energy directly from the sun.

pumper—device for sucking oil out of water.

sedimentary rock—rock formed beneath the water, usually at the mouths of rivers, from material that settles in the water over long periods of time.

skimmer—device for sucking oil out of water.

supertanker—ship designed to carry 200,000 tons or more of oil.

RECOMMENDED READING

Anderson, Madelyn Klein, *Oil in Troubled Waters,* New York: Vanguard Press, 1983.

Anderson, Madelyn Klein, *Oil Spills,* New York: Franklin Watts, 1990.

Brown, Joseph E., *Oil Spills: Danger in the Sea,* New York: Dodd, Mead & Co., 1978.

Lynch, Michael, *How Oil Rigs Are Made,* New York: Facts on File, 1985.

Potter, Jeffrey, *Disaster by Oil,* New York: Macmillan, 1973.

INDEX

ABOUT THE AUTHOR

Christopher Lampton is a free-lance writer. Born in Brownsville, Texas, he has a bachelor of arts degree in radio, television, and film from the University of Maryland.

Mr. Lampton has over fifty nonfiction science books to his credit and nine works of fiction, including several science-fiction novels for Doubleday and Laser Books. He currently lives in Maryland.